Realities Behind Aircraft

BY

SALMAN SADI

Realities Behind Vimana Aircraft

Disclaimer

This book has been written to provide information to help you about the Realities behind Vimana Aircraft. Every effort has been made to make this information as complete and accurate as possible. However, there may be mistakes in typography or content.

Also, this book contains information about Realities behind Vimana Aircraft only up to the publishing date. Therefore, this book should be used as a guide – not as the ultimate source of information on the topic.

The purpose of this book is to educate. The author does not warrant that the information contained in this report is fully complete and shall not be responsible for any errors or omissions.

The author shall have neither liability nor responsibility to any person or entity with respect to any loss or damage caused or alleged to be caused directly or indirectly by this book.

Copyright © 2019 - Salman Sadi

All rights are reserved. No part of this report may be reproduced or transmitted in any form without the written permission of the author.

Realities Behind Vimana Aircraft

Table of Contents

Vimana – The Ancient Indian Aircraft 5

The Rama Empire And Vimanas 6

Ancient Astronauts Theory And Alien Presence on the Planet 7

The Physical Evidence of Alien Existence 10

Vimanas 16

Mention of Vimanas In the Mahabharata 18

Is There A Link Between Vimanas And The Alien Technology? 20

A Look Into The UFO Technology And The Possible Traits Of The Oldest Aircraft 21

Insights Into ADIFO Flying Saucer 23

How Does ADIFO Work? 25

Is ADIFO Going to Replace Airplane? 31

Traits Noticed In the UFOs 32

    1.   Anti-gravity Lift 32

    2.   Sudden & Instantaneous Acceleration 33

    3.   Hypersonic Velocities With No Signatures 34

    4.   Low Cloaking 34

    5.   Trans-medium Travel 34

Heard of The 5,000 Old Mystery Craft In Afghanistan? It's Probably in Pakistan 35

The sudden interest of world leaders 37

The story of the ancient craft 39

Stuck in Time Well 42

How the ancient writings described Vimanas 44

Realities Behind Vimana Aircraft

What happened to the Vimana discovered in Afghanistan? 46

Aliens and Vimana – Is There Any Connection? 49

The Vimana Technology 50

The Nuclear War of The Drona Parva 54

Destruction of Harappa and Mohenjo Daro That's Unexplained To Date 57

Unit (Section) 2: 63

Significance of life organization 63

A Look Into The Vimana Secrets 74

The Vimana Mystery 76

The Atlanteans And The Vimanas 78

The Unexplained Destruction of Harappa and Mohenjo Daro 81

The Possibility of Atomic Wars 83

Unit (Section) 1:

Realities Behind Vimana Aircraft

Vimana – The Ancient Indian Aircraft

There has been a lot of research over the years about UFOs and the alien presence or visits to the planet Earth. But many researchers often overlook an important fact. Although, one may assume that flying saucers have an alien origin, thus, ancient India and the Atlantis remain another possible UFO origin. There have been a lot of ancient Indian sources that give us plenty of information about an ancient aircraft origin. Discovery of written texts and scriptures, "Sanskrit", which passed through centuries is the most reliable scriptures of their times. The authenticity of most of the texts that hint towards Vimana aircraft cannot be questioned. In fact, many of these are well-known Indian Epics.

Vimana aircraft are mythological flying chariots that have been described in these Sanskrit epics and Hindu texts. The Pushpaka Vimana that is believed to have belonged to King Ravana is probably the most frequently quoted evidence mentioned in these scriptures. Even many Jain texts and scriptures have mentioned Vimana aircraft.

The Rama Empire And Vimanas

The "Rama Empire" which was developed some fifteen (15) thousand years back in the Indian sub-continent comprised of the Northern areas of India and Pakistan. The empire consisted of many sophisticated and large cities and many of these cities are still found in deserts, and mountains of northern, western, of Pakistan and India. Apparently, the Rama Empire existed parallel to Atlantean civilization which is located somewhere in the middle of the Atlantic Ocean enlightened by Priest-Kings who ruled this empire. They governed the seven great capitals of the Rama Empire that most classical Hindu scriptures refer to as "The Seven Rishi Cities".

The ancient Indian scriptures refer to these people from the seven rishi cities who were in possession of flying machines and they named them the Vimanas. There is also an ancient Indian epic that portrays Vimana as a circular, double-deck aircraft with a dome and portholes – similar- to what someone would imagine as flying saucers.

The Vimanas had a melodious sound to produce and traveled with the speed of the wind. At that time and age, four types of such aircraft existed. Some of these looked like the saucers while others were shaped in long cylindrical form. The Indian texts that talk about Vimanas are numerous and they had a lot of different things to say about these aircraft. The manufacturers of these aircraft from the ancient times wrote a complete flight manual on how to control different Vimanas. Such manuals can be found in the scripture. In fact, some of them have been translated into the English language as well.

Ancient Astronauts Theory And Alien Presence on the Planet

The researchers have found quite a lot of evidence related to alien experiences from the past in different locations on

Realities Behind Vimana Aircraft

the planet. On some occasions, they believe that the aliens just appeared strangely out of nowhere.

A-lot of us believe in the existence of alien creatures whether they were those ancient visitors or extraterrestrials of the modern-day who keep visiting the planet with some agenda. The creation myths related to all the ancient civilizations discuss a thing or two about alien gods descending from the skies for a range of different reasons. A few of them have been alleged to have mated with the human women for creating bloodlines, while others are believed to create humans with different biogenetic experiments.

The ancient aliens were found all over the planet and there's been enough physical evidence of such out of worldly beings. They don't really refer to the extraterrestrial creatures and rather these were the aliens who make up the consciousness hologram where we explicitly experience and learn. This consciousness is created for studying emotions. While we continue to find out the truth about the creation of human beings and other different forms of life that exist today, we still look at the creation ,"god", of ancient astronauts who landed here from the other solar stellar .

One should not forget that even if the extraterrestrials exist as physical beings, the creation keeps on going beyond any agendas. They are more of a sub-routine within a programmed reality if anything. Now, whether you call it a Master Plan, or a Grand Design, or whatever else you like, remember that it did have a beginning and it is approaching the end at a rapid pace. It is hard to neglect the extraterrestrial signs – that are visible in a physical form.

Realities Behind Vimana Aircraft

The Physical Evidence of Alien Existence

Here we have some pictographs that signify alien interactions from the past.

Sumerian God Anu

He is probably holding the mythical Holy Grail cup

Sumerian Artifacts in the British Museum

Realities Behind Vimana Aircraft

Zecharia Sitchin – The Anunnaki Creation Theory

Is it true that the human race was seeded by these ancient astronauts? What could have been their possible agenda then? Were they using our planet for some sort of 'science projects' they were undertaking, making new races one

Realities Behind Vimana Aircraft

after the other through biogenetic manipulation, and then wiping it all clean before starting again from scratch?

The theorists – who have spent decades researching this topic – believe that the extra-terrestrials did land on our plant hundreds of thousands of years back and they had far superior knowledge about science and engineering. The theories project extraterrestrial intentions for sharing their expertise and knowledge with the early human civilizations to help them change the course of entire human history forever. Even today, the researchers are seeking more and more evidence that could support the prior research.

Realities Behind Vimana Aircraft

This alien theory from the past found its basis in the old idea of existence of life on different planets. It also finds its roots that humans have had alien interaction before as well. It was in 1960s when this idea of human interaction with the aliens was caught in the spotlight as there were quite a lot of UFO sightings. It had much to do with our space program as well because if humans could make it to other planets, then why extraterrestrials couldn't land on Earth?

In the 1968 best seller Swiss author and researcher Erich von DŠniken – Chariots of the gods – he came up with a hypothesis that most of the ancient civilizations acquired

Realities Behind Vimana Aircraft

their religions and the technologies from space travelers. He believed that these old civilizations would have welcomed these extraterrestrials as gods. After his publication people regarded DŠniken as father of the ancient astronaut theory.

The evidences that have been cited by Von DŠniken, may be described as follows.

Artifacts and structures have been revealed representing far superior technological knowledge compared to what could have presumably existed in the ancient times when these structures were manufactured. According to DŠniken, it was either these extraterrestrial visitors who produced the artifacts and/or it was humans who built these after obtaining necessary knowledge from their alien gods.

Some of these artifacts are Moai of Easter Island, Egyptian Pyramids, Stonehenge and submerged  Cities that date back to thousands of years.

Realities Behind Vimana Aircraft

The medieval map that is popularly known as Piri Reis Map and is supposedly the view of the planet Earth from space.

Nazca Lines in Peru which, according to von DŠniken, served as the airfield or landing strips.

There has also been ancient artwork which have been interpreted around the world as possible depictions of the ancient astronauts, space vehicles, complex technology, and the extraterrestrials.

DŠniken's evidence also include the explanations of new religions originating after human-alien interactions and he has also interpreted Bible's Old Testament. According to him, ancient humans thought aliens' technology to be far

Realities Behind Vimana Aircraft

superior and supernatural and they believed them to be the gods. DŠniken questions if the literal and oral traditions of most of the religions have references to foreign visitors from stars as well as the vehicles that travel through space and air.

Vimanas

Finally, the Vimanas, according to DŠniken, should be taken as the literal descriptions that have changed a lot over the years becoming much more obscure. Examples include:

- the revelation of Ezekiel in the Old Testament that describes in detail the landing of an aircraft full of angels who looked more like man.
- Moses and the God's directions to him about building Ark of the Covenant that, supposedly,

Realities Behind Vimana Aircraft

serves as a device for communicating with the aliens.

- The extended family of Lot who received the orders from human-like 'angels' to get to the mountains because the destruction of the Sodom city by God had already been planned. His wife tried to look back at nuclear explosion that fell "right where it was intended to". DSniken tries to highlight similarities with "cargo cults" which formed in World War II and afterwards where the once-isolated tribes from South Pacific wrongly assumed the advanced Japanese and American soldiers as gods.

Realities Behind Vimana Aircraft

So, if the aliens had interactions with the humans on Earth, could they return at any time in future? According to the ancient alien theorists, they surely can. According to them, these aliens share their views and the technology with our world in preparing our future generations for such inevitable encounters that are bound to happen.

Mention of Vimanas In the Mahabharata

Mahabharata, an ancient epic in Sanskrit dating back to almost 800-900 BC, also mentions Vimanas and describes them as the airborne chariots powered by "the winged lightning". The reference of the Vimanas shows that the purpose of flying machines is not just traveling on the planet, hence, the purpose was also space travel. The reference of Vimanas presents this fact that this ancient craft was able to soar, 'to both the solar and the stellar regions.

The 164.47-48 verses in Mahabharata demonstrate further evidence of this idea of the vimana aircraft that are often

Realities Behind Vimana Aircraft

referred to as mechanical birds. The verses describe them as something of dark descent and golden in color that goes up into the heavens and then descend back to the Earth. They have twelve fellies with a single wheel and three naves. A total of 360 spokes are fixed into the fellies.

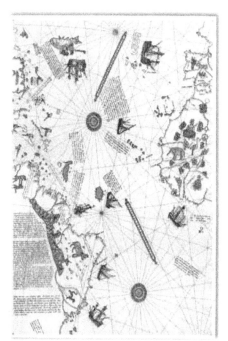

The translation of Swami Dayananda explains these verses as "speedily jumping into space with a spacecraft using water and fire...containing a total of 12 stamghas, 3 machines, 1 wheel, and 300 pivots besides 60 instruments".

Here is how it describes the Pushpaka Vimana of Ravana: "My brother owns the Pushpaka Vimana which resembles the Sun and the powerful Ravana, and aerial Vimana can go anywhere at will….the chariot resembles bright clouds in sky…and as the King entered, the excellent Vimana rose up and went into higher atmosphere at Raghira's command".

In years to follow, the Indians started building temples which were formed in the shape of these Vimanas just as described in the sacred texts. The buildings resemble a lot to the rocket ships of the modern day. In fact, they represent the physical evidence of alien technology lost long in the past.

Is There A Link Between Vimanas And The Alien Technology?

Well, all these debates above suggests that humans and aliens had interactions and the theory that says the aliens were far superior in technology than humans at that time establishes the link between the vimanas and the alien technology. Even when the vimanas and mechanical birds were mentioned in the Hindu scriptures, that vimana's very existence could not have been possible unless there was some alien technology was in place to make it all possible. Ancient astronauts and the researchers are coming up with similar theories and even physical evidence supports such theories, the vimanas are bound to have some link with the alien technology otherwise such mechanical master class could not have been possible in times immemorial.

So, what do you think of these Vimana Aircraft from hundreds of thousands of years ago? Would you like to go with the same ancient astronauts' theory? Or you have your own take on these ancient aircraft? Decide for yourself!

A Look Into The UFO Technology And The Possible Traits Of The Oldest Aircraft

The technological boom of this century can be referred to the UFO technology. Let's see how it lives up to the rules of modern physics. Is it possible to come up with an explanation about those oldest aircraft by closely

Realities Behind Vimana Aircraft

examining the credible UFO reports available today? Yes, some believe that it is very much possible, and they have even come up with their explanations of the phenomenon as well.

There is one idea commonly represented in some of the credible UFO reports available to us as they all represent some very advanced aircraft with extraterrestrial origins. However, it's been in the recent future that modern physics has been able to come up with theories that directly point towards a possible explanation of the alien technology used in these oldest aircraft.

Today's modern physics tries to combine all of nature's fundamental forces "Theory of Everything" and proposes theories which call for somewhere around 11 space-time dimensions to be existed along with a possible existence of entire universe alongside one where we all live. Thus, doesn't just show us the evolution of physics to this very point but also tries to find out some of the oldest unanswered questions from modern physics like origin of

Realities Behind Vimana Aircraft

inertia and the mystery of comparative weakness of gravity.

These unanswered mysteries, as well as some speculation and pertinent characteristics of UFOs, can be combined with the modern-day physics theories to form the possible explanation of the UFO technology in today's tech-advanced era.

Insights Into ADIFO Flying Saucer

Even though we humans have reported sightings of the extra-terrestrial flying saucers for ages, it appears that in

Realities Behind Vimana Aircraft

2019 we have come so close to understanding the UFO technology that we have some verifiable documentation about their existence. We have the modern-day flying saucer that goes by the name ADIFO.

Even though it's been created by the terrestrial hands and belongs to Razvan Sabie – an engineer from Romania – this flying saucer that goes by the name ADIFO or All-Directional Flying Object – has been able to capture the attention of those who are interested in the UFO technology. One key reason behind that may be is the semblance of this terrestrial flying object with what we have grown up imaging an extra-terrestrial craft alike. According to the news sources from Romania, this ADIFO flying saucer has been formed in this shape to purpose this vehicle for some hyper-maneuverability. According to the papers, the inventor states that the device is one of its kind with the ability to evolve in all directions without any change in its aerodynamic characteristics. It is

Realities Behind Vimana Aircraft

even capable of flying equally well in supersonic or subsonic regime.

This test model from the Romanian engineer appears to have a promising future ahead as it comes as a radical invention among recent flight technologies and has already received international press award as well as a gold medal from 46th international inventions exhibition in Geneva.

How Does ADIFO Work?

Giving this flying object the shape of a saucer wasn't just some cheap gimmick. Unlike the typical aircraft with wings, which can only make linear movements, this saucer form allows the craft to have maximum agility.

Realities Behind Vimana Aircraft

It shares a few similarities to drone aircraft that hover in the air with the help of 4 ducted fans and is still able to assume the aerodynamic capacities of high-speed jets courtesy with two (2) thrust nozzles in its lateral part. These thrust nozzles closely resemble the ones you may find on spaceships.

Furthermore, this ADIFO aircraft is designed in a particular manner so that it can avoid sonic boom which shows up when jets break speed of sound. It is able-to-do so with the help of its distinct form that makes a cleaner transition possible through various velocities.

When the speed is low, ADIFO works more like a simple quadcopter while it is like jet-propelled and extremely efficient supersonic craft at a high speed where its entire body works like low-drag wings. According to the claims of the Romanian engineer, the saucer has been designed to provide unparalleled aerial agility over a spectrum of speeds.

Realities Behind Vimana Aircraft

To put it simply, ADIFO looks like a disc in its shape and its entire surface works as the wings. To be very specific, it's been shaped in a way that it mimics dolphin's airfoil back half of shape and can radiate out in every possible direction from its center. Its outer edge tapers to thin ring which makes it slippery when flying horizontally.

Slow speed and VTOL maneuvers are performed with the help of four(4) ducted fans that allow the ADIFO saucer to work like a typical quadcopter drone. It also has a thrust nozzles on its rear end which are meant to offer a horizontal thrust and can also vector independently to get high amount of agility during level flight. When moving at a high speed, smaller discs show up and cover up the fans of the quadcopter to ensure an even better profile. Similarly, the legs have the ability to retract as well.

The eventual propulsive touch comes from a thrust-nozzles placed laterally and point out towards each side. They work more like the thruster systems for reaction control in spacecrafts. Allowing *ADIFO* to take a

Realities Behind Vimana Aircraft

horizontal flight, with a similar characteristic to flying saucer, pushing itself rapidly sideways in each direction. It even allows the flying object to rotate rapidly during flight. According to the creator, it lends all the maneuvering capabilities to the saucer that are simply unmatched. And, it doesn't even need separate wings, flaps, rudders, or ailerons to perform maneuvers.

In fact, there is more to it. The flying saucer can also fly upside down, both in the horizontal flight and the quad mode. When provided with correct jet propulsion, it turns out to be efficient with supersonic and transonic speed. Besides, the modeling the ADIFO team suggests that there won't be any typical sonic boom.

Realities Behind Vimana Aircraft

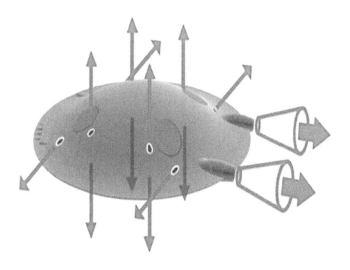

Even though this prototype is completely radio-controlled and an unmanned aircraft, the creators claim that they can democratize its supersonic flight if built into a multi- or single-seat manned aircraft that uses hybrid jet/electric propulsion system. It's, however, going to be quite interesting to see how this team can build the pilot visibility and what kind of controls will be needed to manage the flying object's various control options and flight modes.

Realities Behind Vimana Aircraft

It is quite a fascinating idea and it might offer the mind-boggling acrobatic flying capabilities to the flying saucer when ADIFO is introduced in its final version. It really replicates the idea of a UFO in every respect from its design to the technology that could have been used in those flying objects at that time. Nothing else could probably dart about and hover like drones while also providing that extremely perfect high-speed performance. There's no other flight technology that could provide such spinning ability and/or a sudden thrust at speed in five(5) different directions. In fact, the main ducted fans can even flip or tilt this craft in the horizontal flight as well. Yes, that's really something special and mind-boggling imagining such craft could be operated by a trained pilot as well in near future.

It doesn't even feel like something that is ludicrously far-fetched to be built. We already have quite a few electric multirotor being developed that will work as manned machines and will have almost similar kind of features as this ADIFO aircraft promises, thus, in the low-speed flight. It's without any doubt that all these things are actually- happening today. Those vectored thrusters used at the saucer's back end aren't something new and same is the case with jet propulsion which is getting quite common and even more reliable. The tapered shape of the body doesn't seem like something impossible either. In future the models of ADIFO may contain more ducted fans, for redundancy.

Is ADIFO Going to Replace Airplane?

Looking at it as standard engineering and science view, it is hard to predict ADIFO replacing Airplanes. The creators, however, are hopeful that they will be testing out ADIFO's full-sized models with pilots on board. Besides, they believe that their invention is appropriate for both civil and military use. So, the next time if someone reports the sighting of the flying saucer, they might be referencing to an ADIFO rather than a UFO as they both look similar and may even have the similar technology used in them. Something that once used to be quite imaginable as some alien invention of the future now has become the reality.

Traits Noticed In the UFOs

Now that we have enough of the modern technology from the Romanian creators of ADIFO, let's take a look at some of the traits that used to be observed and reported commonly in the UFOs. We'll surely then try to develop some type of link between how similar these two kinds of aircraft are comparable.

"Let's first find out the traits that have been reported for UFOs to have".

1. Anti-gravity Lift

Quite dissimilar to the common aircraft that you might have known until today, the UFOs used to overcome the gravity of Earth without any visible propulsion means that they didn't have any kind of flying surfaces like wings either. Refer to the Nimitz incident, for instance, where the witnesses described that the crafts, they saw were tubular and were shaped similar to some Tic-Tac candy.

2.   Sudden & Instantaneous Acceleration

The UFOs were objects that could change their direction or accelerate very quickly, and the magnitude was such that the human pilots couldn't even survive those G-forces they would have experienced in such events. The radar operators report about Nimitz incident that they tracked a UFO which was dropping from above at a speed that was almost 30 times greater than speed of sound. A fighter pilot who was sent in to intercept a flying object observed the object moving at rapid speed side-to-side movements and which were captured later-on in the infrared video. The object looked like a ping-pong ball according to radar operators, the object's acceleration went from rest-0 to 60 miles within a minute – it's an amazing 3600 miles per hour.

Realities Behind Vimana Aircraft

3. Hypersonic Velocities With No Signatures

When some of the aircraft moves at a speed faster than the sound waves, it usually leaves some signatures – such as the sonic booms and vapor trails. But this was not the case with UFOs and many of the UFO encounters we have had to this date did not leave behind similar characteristics.

4. Low Cloaking

Normally, when you are observing different objects, it is really hard to get a detailed and clear view of these objects either through radar, pilot sightings, or any other means. Generally, the witnesses are only able to see the haze or glow around them and what exactly used to be the case with UFOs.

5. Trans-medium Travel

We have seen a few UAPs moving easily across various environments, like space, the atmosphere of the Earth and even in the water. In-reference to Nimitz incident, the witnesses saw a flying object hovering above a churning "disturbance" right beneath the ocean's surface. This led to the speculations that some other UFO might have went through the water. A radar operator confirmed it later through Navy sonar operator that there was a craft moving quicker than 70 knots, which is approximately twice as fast as speed of the nuclear subs.

So, if you look at most at these observed traits of the UFOs and compare them with all the details shared about ADIFO, it is evident that the older crafts, "such as vimana", used to have similar technological characteristics and exhibited similar traits.

Heard of The 5,000 Old Mystery Craft In Afghanistan? It's Probably in Pakistan

The world is filled with unsolved mysteries people are still trying to figure out. Before we could be able to find extraterrestrial life outside Earth, we need to solve mysterious encounters here. There are many incredible and inexplicable situations on our planet that governments keep away from us and/or simply cannot be explained yet. Imagine how many mysteries the secret intelligence

Realities Behind Vimana Aircraft

services in each country hold. So, it's no wonder that when an incredible, unexplainable situation flashes on the news, people are all over it and start to make out theories, find out as many details as possible and contextualize it until there is no news left.

One very interesting, and certainly discussed, topic is the mystery craft found in Afghanistan couple of years back – a situation that was not made clear until today.

Back in 2012, on a normal day, eight soldiers from the United States made a very strange discovery while they were in Afghanistan. They were checking out an area in the country, clearing it out from terrorism when they made a strange discovery in a cave nearby. An ancient unknown object was set before their eyes, an apparition they could not explain or know how to handle at the moment. In just a few days, the news traveled all over the world and many important political leaders started expressing interest in this uncanny discovery. However, until this day, many question marks had arisen.

The story is that the strange craft that had been first seen by the American soldiers was, in fact, an old Vimana, known in the Indian culture as a vehicle to travel through the air. Moreover, the aerial object was no less than 5,000 years old, making it more interesting to people around the planet. Once the odd craft had been revealed, many tried to solve the mystery behind it. Did they succeed? Read on to find out more interesting information on the topic.

The sudden interest of world leaders

The odd visits of all world leaders to Afghanistan happened before the news about the craft was even revealed to the world. The conclusion of this Odd mobilization of world leaders in Afghanistan raised suspicions; that what is happening in Afghanistan that made everyone so interested? Because In just a few days, such deployment of the world's most important people in Afghanistan was more than a standard meeting.

Many famous world leaders took a sudden interest in this region. Within a week, important leaders at the time made short visits to Afghanistan which made the world suspicious. One after another, names like David Cameron, Barack Obama, Angela Merkel, and Nicholas Sarkozy all visited the scene. The question is what was so important that some were eager to conventional travel from their offices, while others were involved in other official visits (for example, Nicholas Sarkozy suddenly left India to fly to Afghanistan). This made the world wonder what is behind this story. Soon after that, the United States leaked an answer – a very old craft has been found in an Afghanistan cave. The ancient object was even more

Realities Behind Vimana Aircraft

interesting, as it was believed to be a flying machine, called Vimana. The question is if this information leakage was intentional, or reverse psychology? because the government is known to hide all extraterrestrial contact. What do you think?

The story of the ancient craft

The word Vimana, which was given to the strange ancient artifact discovered by the American soldiers in Afghanistan, has multiple meanings in the Indian language. Vimana is both a palace, a temple or a mythological flying machine. It probably depends a lot on the context. However, this language barrier only increased the mystery around the ancient object and still raises many questions.

The Vimana in our case is thought to be an ancient flying machine, referenced in multiple Indian sources. One of the most important sources is the Indian Epics, an ancient cultural heritage of India, where many such artifacts are described and used. The most curious thing is that a large portion of the Indian Epics has not been yet translated into English, as the world only has the original writings in old Sanskrit.

The ancient craft discovered in 2012, in Afghanistan was similar to the old Vimanas described in the Indian Epics, having 12 cubits circumference and 4 strong wheels. It is believed that these were deadly weapons used for destruction. A Vimana of this kind is used with the help of a circular reflector that produces a powerful light when it is switched on. This light can focus on any type of target that will be immediately destroyed, thanks to the craft's power.

In the same cave where the old craft was discovered, some writings were uncovered too. The writing on the wall may possibly be referring to the machine belonging to Zoroaster, a well-known ancient prophet and also the founder of Zoroastrianism.

Realities Behind Vimana Aircraft

After the ancient object was discovered in the cave, people rapidly understood that it was more than an old stone. It has been said that the ancient craft is protected by a very powerful energy barrier that prevented anyone to remove it from the original place where it was found. At first, it was thought that the eight soldiers that originally discovered the ancient craft went missing. After the incident happened, the United States military sent scientists to investigate the  discovery on the ground. However, the soldiers were later found, after they supposedly had some good rest from tough hike, "really".

Realities Behind Vimana Aircraft

The mystery that surrounds this ancient craft hasn't been solved until this day. American scientists also inspected the place, hoping to find more information about the Vimana and the causes that kept it inside the cave. They believed that the ancient craft was trapped there is a "time well". The ancient ship seems to be large enough to fit a sizeable number of people. Some commented that the ancient craft was as large as a city!

Stuck in Time Well

Vimana was trapped in a time well circulated all over the world. The time-well is a term made popular by Albert Einstein in the "Unified Field Theory". It is believed to be an electromagnetic field. The term also appears in the theories surrounding the World War II –it is believed that Americans made a teleportation experiment in 1943 known as, "Philadelphia Experiment", that was based Einstein's- Unified Field Theory.

Realities Behind Vimana Aircraft

Furthermore, there are many references with explanations, followed by energetic solutions(ES) and construction designs(SD) in multiple Hindu, "Sanskrit", writings. One of the most important Hindu writings talks about ES/SD in Vimanika Shastra. There it has been stated that "a" Vimana-the flying chariot can fly at superspeed limits on the terrestrial atmosphere, and under water and/or through the oceans. Nevertheless, the Vimana can also fly through outer space stellar. Some people studying the phenomenon believed that the Vimana could have also been a death ray or nuclear bomb.

Another theory surrounding the discovery of the old ancient craft is that it's not the first one of this kind discovered so far. In-fact, a few years back, some documents in Sanskrit have been discovered in China that described how to build an interstellar spaceship, similar to Vimana. The documents have been sent to India and translated there, igniting, the facts, that our ancient civilizations were, in some perspective, more modern than we are right now! They believed in an anti-gravitational method of propulsion. They used an analogous system for the lightmap ("the unknown power of the ego in man's physiological makeup") to build these ancient spaceships. In the texts discovered in China, the machines were called "Astras" and the people of the time believed they could have succeeded in sending men to another planet.

How the ancient writings described Vimanas

Technologies of Vimana crafts are detailed in many Indian ancient writings. For example, in the Yujurveda, there is a tale of a flying machine that has been used by the heavenly twins, Asvins. Vimana means flying machine, being a synonym for the phrase. The word occurs in multiple writings; the Ramayana, the Yajurveda, the Bhagavata Purana, the Mahabharata and in the Indian literature.

From all the ancient writings, there were 20 passages in the "Rigveda" that refer to a flying vehicle of the Asvins.

Realities Behind Vimana Aircraft

The flying machine described is triangular, it has three wheels and is three-storied. The spacecraft carries at least three people and is made by iron, silver and gold, with two wings. The story says that the Asvins used the flying machine to save King Bhujyu, who was at the moment in distress on the sea.

The majority of scholars today know the "Vaimanika Shastra", which is a sketch collection. The center character of this is Bharatvaj the Wise, who lived somewhere around 4th century before Christ. These writings were found again in 1875. Most of the text describes important parts of flying machines and their sizes. From these, anybody can learn how to steer flying machines, what precautions to take on a longer flight, how to protect the machine in dangerous situations, such as a storm or lighting. Also, in the writings it is described how to land if forced by the situation or how to use the solar energy to drive and save fuel. Imagine that there are no less than 70 authorities and ten air travel experts referred

Realities Behind Vimana Aircraft

in Bharatvaj! Plus, the very accurate descriptions of the old flying machines made contemporary experts amaze. There are still a few language barriers – for example, some metals from which the spaceships were made of cannot be translated yet in English, making it unclear for us what they supposedly used when building Vimanas. In these writings, there were four types of flying Vimanas: Sundara, Rukma, Skuna and Tripura. It seems that the Indian ancient culture really focused on travelling by air and building crafts that will not only help them travel but protect them as well. Until the day, with all the information we have from the ancient texts, there are still many things that remain unclear.

What happened to the Vimana discovered in Afghanistan?

After the massive attention the incident received from the press and from the global leaders at the time, the news faded, and nobody talked about it anymore. There are some articles that are written on the topic from time to

Realities Behind Vimana Aircraft

time, but much of the attention that sparkled when the news first appeared has gone away. Why? Maybe because there is more to the story than what we are led on to believe.

Some believe that the entire story was just an invention to put that part of the world on the map and add some extra attention on Afghanistan. Some believe that the incident never happened. However, the majority of people think the ancient space craft exists, but a-part of the story has been hidden from the large audiences.

Right now, the ancient craft might be hidden somewhere in the world, close to the place where it was found, but away from the time well, so it can be studied better. However, this would not explain how the ancient craft was moved and how people successfully got rid of the energy barrier mentioned earlier. Some believe the ancient craft is hidden somewhere in Pakistan, in a place only known by the Pakistani Intelligence. Unfortunately, little is known about the current location of the ancient craft, or if it even

Realities Behind Vimana Aircraft

exists. Perhaps, maybe Pakistani Intelligence is hiding its' location, most likely, will never be shared with the world. In conclusion, the mystery remains unsolved to this day. With all the theories surrounding this topic, it is difficult for the large audience to see the clear truth. There are many details hidden by us. Maybe there are so many hidden  things to protect us or to keep us in a comfortable dark. We don't know. But what we do know now is that a 5,000 years old craft has been discovered a few years ago in Afghanistan and the news disappeared as fast as it appeared. What happened to ancient artifact? Was it even true? Will more details leak in the years to follow? Is the

Realities Behind Vimana Aircraft

Pakistani Intelligence really hiding it from everyone? What do you think?

Aliens and Vimana – Is There Any Connection?

In reference to Hindu scriptures, they're full of amazing stories about Hindu gods, the powers they had, and the epic battles which supposedly happened long time ago. Mostly, their sagas are perceived as nothing but mythological stories that had been carved out for being taken as a metaphor, just the way we tell different fables to our children so that they can pick up some useful lessons for the life and apply them as they move forward. The myths from the Hindu scriptures talk about their noble gods fighting off some wicked forces, and flying ancient aircraft known as Vimanas. Some of these epics even have mentions of massive wars, maybe a nuclear war. Some say they were Hindu gods; others say they were aliens. Are we ever going to find the real-truth? Are these nothing but allegories? Let's try and dig a bit deeper.

Realities Behind Vimana Aircraft

The Vimana Technology

Let's first try and find out some details about the Vimana crafts and the technology they used. The ancient Hindu scriptures describe these crafts as flying machines with varying degrees. When you translate the word Vimana from Sanskrit, it translates to "traversing" or "something that's been measured out". The texts say that these machines were flown by the gods themselves in some sort of wars for the better good. Quite similar to the chariots discussed in these biblical texts, particularly the one that's seen in the Ezekeil's vision of wheel, such flying crafts were present in all sizes and shapes and they had the capability to travel at various speeds and to different distances. Some of these used to be seafaring and land vehicles,whereas, others had the capability to fly, and some could even go into the space.

The most prominent of all texts about these ancient Vimanas come from popular Vimanika Sastra, a translation from early 20th century that described different

Realities Behind Vimana Aircraft

aspects of the Vimana technology from oldest Vedic scriptures. There are a variety of drawings of different crafts, including fuel sources that were used to supply them with power. However, some of these can be quite confusing. These translations describe various elements as well as minerals that we're familiar with today, such as quicksilver, mica, and mercury. Besides, there are mentions of some strange liquids which were mentioned as honey but could be some unknown substance having somewhat similar viscosity as the nectar of bees.

RUKMA VIMANA

VERTICAL SECTION

Drawn by
T. K. ELLAPPA,
Bangalore.
2-12-1923.

Prepared under instruction of
Pandit SUBBARAYA SASTRY,
of Anekal, Bangalore

Realities Behind Vimana Aircraft

These Vimanas could be found above all the Hindu pyramids and temples. Usually, they were round and saucer-like objects that are believed by some theorists to be extraterrestrial vehicles. That's where the links of these chariots to aliens could have been established. According to Erich von DŠniken, the sightings of the flying saucers in modern times which actually shaped our perception about the UFOs are similar to these Vimanas from ancient India.

DŠniken has also pointed out towards depiction of the Shiva taking a flight on Garuda and believes that it could simply be an ancient description of the spacecraft or an airplane. Garuda, the flying bird of Shiva, was popular for flying into the space and reaching the moon, dropping bombs, and taking Shiva to various locations throughout this solar system of ours. Explaining this type of sight to the future generations, the story that elders told about gods flying around to different locations on a big bird or, maybe, some bird-like aircraft would really sound ridiculous. Everyone would think of it as nothing but some mythological tale if they've never witnessed such-a-thing.

A closer look at the Vimanas reveals that the descriptions about the sounds created by these crafts, and how they apparently looked while taking off, resemble a lot to the jet propulsion. In fact, there are a lot of other similarities as well. Here is a translation of one passage from Vedic Mahabharata which describes Vimana:

"The Vimana was equipped with everything necessary. Even the demons or gods could not conquer it. It also radiated light as well as echoed with deep sound. The beauty of the craft captivated minds of everyone who beheld a Vimana. Visvakarma, the construction and design lord of a Vimana, had made it with his power of austerities. Its outline, similar to that of sun, couldn't be delineated easily."

The passages talk about the cohort of Krishna and Arjuna, the epic hero from Bhagavat Gita, while describing the trip taken by him into the heavens above using a Vimana craft. It was where he witnessed a large number of airborne chariots as well as other huge Vimana which had 7 stories on it and was quite high. Just like the trip Enoch had taken up in some wheeled chariot, Von DŠniken says it could be some primitive interpretation of the trip taken to mothership that could have been the origin for a lot of Vimanas seen on the planet Earth.

The Nuclear War of The Drona Parva

There have been strange stories about massive wars in the Hindu scriptures that also point towards the existence of Vimanas and the aliens or Hindu gods. One such story comes from the translation of Drona Parva which is the 7th book from Mahabharata. This book shares the story of Drona who was a warrior given the responsibility to lead an army during Kurukshetra War. The book also shares the story of Drona's ensuing death during the battle. This story fits perfectly into the themes that can be seen

Realities Behind Vimana Aircraft

everywhere else in Mahabharata, as well as other texts detailing all the difficulties one can face in a war. However, this book shares some descriptions that have some eerie similarities to specific effects of nuclear wars. For instance, it talks about the explosions leveling everything up, animals stuck in the flames and screaming miserably, babies dying inside the wombs of their mothers, and the metal shields melting on the warriors' skins. This all sounds like nothing but the aftereffects of some serious nuclear blast. The book also mentions birds falling off from the skies because of a projectile carrying the complete universe's power, and so bright as thousand suns.

"We beheld as we saw into the sky a scarlet cloud that resembled some ferocious flames of the blazing fire. A lot of blazing missiles seemed to flash out of that mass, and there were tremendous roars, such as noise from a large number of drums all beaten at a time. A lot of weapons having gold wings fell out of it and there were huge

Realities Behind Vimana Aircraft

thunderbolts, loud explosions, and a large-number of fiery wheels."

So, what did these fire-resembling clouds refer to? What did the subsequent destruction and death mentioned in those ancient scriptures describe? Could it be effects of the atomic fallout? The ancient technology from that time could not really have seen the exposure to any kind of radiation. However, descriptions of the babies of pregnant

Realities Behind Vimana Aircraft

mothers dying portraits a aftereffects of radiation exposure.

After first atomic bomb was developed which saw its fate on Nagasaki and Hiroshima, Bhagavat Gita was even quoted by Oppenheimer saying, "Now I am become death, destroyer of the worlds." What is so odd in this entire episode was that Oppenheimer, the maker of the atomic bomb, also knew Sanskrit and was even a scholar in that language, and some even compare Oppenheimer's story to Arjuna's Bhagavat Gita. Even Arjuna must have been convinced to fight a battle that he didn't want to participate in because of some moral dilemma. Some have compared Bhagavat Gita, "though in hesitance", with Oppenheimer's success in development of atomic bomb.

Destruction of Harappa and Mohenjo Daro That's Unexplained To Date

You must have heard of various ancient civilizations from the popular Indus Valley in today's India and Pakistan. There are some remnants of these civilizations available today that have been secured by the archeologists, but they seem to have been puzzled by these remnants for decades. Both cities were deserted mysteriously, and coincidently, it was the time when the great pyramids were being constructed in Egypt. It all happened at the time when these cities were considered the thriving centers of culture and technology for centuries.

Particularly, Mohenjo Daro seems to have flattened out or collapsed somewhere around 600 years after it was first inhabited. The question arises, and many ancient theorists have this viewpoint, that it could have been the location of nuclear wars fought in ancient times between the aliens or the mythical gods discussed in the Hindu scriptures. Harappa and Mohenjo Daro are both anomalous not only because they were deserted by people living in these civilizations, but also a due-to-the fact that they both had inseparable advanced technology at that time and were secular, decentralized societies. The archeologists have figured out the layout in which these cities had been built and inhabited. According to the specialists, their layout shows a kind of an-evidence of egalitarian civilization without any distinct hierarchy. There wasn't any ruling class in these cities, and they were probably governed by the elected officials.

If you have a look at Mohenjo Daro's ruins that are present in modern-day Pakistan and have been preserved over the years, you would notice careful planning of the urban cities, advanced drainage and irrigation systems, and watertight, 900 sq. ft. a communal bath that used to be filled through the water of Indus River. The approximate area covered by this "particular" civilization was 500 acres. It housed somewhere around 20000 to 40000 citizens at that time. An impressive number for ancient times to support the population, especially, considering the careful planning to sustain an infrastructure around

Realities Behind Vimana Aircraft

these cities.

On the other hand, Harappa came into a ruin somewhere around that same timeframe when Mohenjo Daro faced its demise. The civilization was equally advanced as its counterpart and had superstructures, granaries, and some calculated practices of trading. Both civilizations had popular board games such as chess, valued purity, and traded precious jewelry and gems. Both civilizations had thriving cultures that suddenly disappeared and there seems to have no proper explanation for their demise to this date.

Realities Behind Vimana Aircraft

The excavations done at Mohenjo Daro revealed skeletons of various people who are found to be holding hands. As it appears, they've been flattened out with ash and rubble foretells an unforeseen deadly event that happened abruptly. According to some accounts, there must have been a certain layer of some radioactive ash found in the soil of the site. This further elaboration gave strength to the theory of this nuclear event which is believed to have caused the ancient cities to destroy.

However, claims about the destruction of the two cities as a result of nuclear attacks from aliens or the Hindu gods are a little unsubstantiated with not much evidence available to support the discovery of the radioactive material from the ruins. It appears to be quite strange, yet, that the two extremely advanced civilizations went through such apocalyptic events at the same time in history. Even though they were located in the same geographic region, still they were a lot of miles apart from

Realities Behind Vimana Aircraft

each other. So, what could have possibly caused them to vanish from the planet all of a sudden?

Whether the stories of the demise of two cities found in old Vedic scriptures provide some real evidence about a nuclear war in history or not, at least they hint us about some extremely advanced and apocryphal technology. Is it possible that there might have been some flying crafts with the description matching that of Vimanas? Did a nuclear event, actually, happened and wiped out both the ancient civilizations of Harappa and Mohenjo Daro as described in Mahabharata scriptures? And, in case it did happen, could these Hindu "gods" who developed and run this advanced technology have an extraterrestrial origin which is not so hard to believe for many of us today? There are both types of theories and theorists who have their split opinions about the link between vimana and the aliens.

Unit (Section) 2:

Significance of life organization

My research on Vimana led me to explore the desert and northern-Pakistan on multiple occasions. Therefore, my experiences are based on multiple events in northern and southern Pakistan. Now the question is why I chose to search for vimana in northern-Pakistan; well let's intercept the Mahabharat saga. When Pandavas went into exile, according to Sanskrit, portraits their time of banishment in a forest verdant with mountains flourished by the river where the band of Pandavas, "Sandal[Sethi]/Judah/Quraysh", built the Kuru kingdom. Another clue is mentioned in the battle of Kurukshetra that took place in the Kuru region that was facing the rough mountain terrain, lush forests near the longest river which portrays the scenery of northern Pakistan. For the most part, northern Pakistan was not habited by humans for thousands of years before and after the epic of Mahabharat, and even to this day, it is considered the most

Realities Behind Vimana Aircraft

remote place in the world based on human ratio inhabiting land across the globe. Tough terrain, and rough path, extreme cold makes it 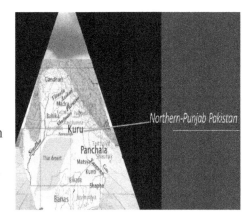 impossible for humans and transport animals to cross the valleys through northern Pakistan.

My journey had many adventures and dilemmas; thus, this expedition was worth all the risks. Nonetheless, if there is ancient hidden technology, I will discover it in Northern Pakistan. My first visit to Pakistan from the airport to the bungalow, I must admit, that on occasion I try convincing myself to turnaround to go back to the United States. The overwhelmed feelings due to cultural shock can play with your mind. For the first-timers, let this be a warning, that it will take some time to adjust to the environment and cultural difference. Though, speaking from experience, I promise when you come out of blinded perception, you will

Realities Behind Vimana Aircraft

appreciate the authentic simply developed culture, moreover, the simplicity of life will fill you with joy. You will get to ride a horse-driven cart," known as tanga," you will get to ride the tricycle-rickshaw and will feel every possible bump on the road. If you are the environmentalist advocate you will be facing your nightmare, for record, only in the densely inner cities. Oh, and I don't want to forget the hospitality of people, where you will be drowned in tea-Chai, "yes", at every stop you will be drinking a free cup of tea, so forget about getting dehydrated, you will be full with a dairy product.

My epical expedition finding the Ancient Vimana aircraft is based on multiple visits to Pakistan. Let's cover general geographical areas that I was able to reach and most likely will be reaching other parts. Lets' cover the cardinal direction, where I believe the saga of Mahabharat took place between North-west and North-East area of Pakistan, $\underline{270°}$ $^{NW}\underline{360°}$ $^{NE}\underline{90°}$ surrounding the Indus river, I have covered a small chunk of the area so far where I think the ancient craft,

Realities Behind Vimana Aircraft

is possibly hidden. First, my search started from the roots of understanding the time of the Sanskrit portrait empire from the Jagannath temple in Sialkot, Qila Kalarwala-Hindu temple in Sialkot, Sun temple in Multan, and multiple mandirs," Hindu temple," in Islamabad. After studying the temples, I found substantial clues directing, near the Indus river and between $\underline{270°\,^{NW}360°}$ $^{NE}\underline{90°}$. So, I began

my search around the Indus river; I must say it was an extreme hike, along the river.

I have discovered many hidden pathways, "leading to small valleys", and discover multiple ancient ruins which I will save for later research and most likely revisit to find out what exactly I was looking at. The next journey I took

Realities Behind Vimana Aircraft

was remote edges of Suki Sar, and the best way I can describe Sukai Sar is going to Northern Michigan, Northern Wisconsin, North Dakota, Colorado, and the Alaska United States. If you have visited all these states you will feel the same around Suki Sar, accept the luxury of technological comfort, if unprepared you will be concealed by the ancient Kuru Kingdom, according to my accounts, being prepared is essential to survival. As I explored Indus river, I've also started to explore the areas around Sukai Sar, I remember being lost many times, not sure how much distance I covered, but I found myself in very remote places where finding another human would be miraculous. As I penetrated deeper into snowy hills, lush jungles, my presence promoted the awakening of the Kuru Kingdom. At some point I was bestowed between time, I had no idea about my whereabouts, however, as I regain my directions, I found myself surrounded by extreme silence, as if, the time had stopped. Before, panicking I took a deep breath, and here comes the sound, that is hard to describe, it sounded like a Jet engine, hence, a dense sound, similar-to the ship horn. This was the sound that it felt familiar, though, I have never experienced hearing it before. Suddenly, I was drowned by the fear-fury, to overcome I convinced myself, that what I'm hearing is perfectly normal. While fighting weak feelings, I thought about various scenarios that possibly led to this extraterrestrial sound. Perhaps, it could be a Pakistani air force conducting flight training. But then how come I was not made aware during my arrival in Pakistan to avoid such areas. Truth, to the matter, is that I was fighting myself thinking if it is not the sound of a fighter Jet, then what am I hearing. After some time, I came to my terms; I remember my goals, remembered what brought me here and remembered that I had accepted all the risks before coming here. Then said to myself in a raised voice, discovery requires risk-taking, by overcoming fears and facing danger, so I shall move forward, with a prohibition to whatever is coming on my way. This rumble gave me

Realities Behind Vimana Aircraft

the will to regain my old adventures self again. I heard the extraterrestrial sound again, but the echo of the sound had no directions, as it elongated. Echo draw me out further deep into the valleys closer to where the sound possibly was generating.

I found myself facing the mountain, thinking about how I got here, though, looking at the mountain, it would measure to a hundred-story apartment building size. The sound produced by the object coming from inside the mountain seemed to be configured to on and off triad because it would take approximately 25-seconds before you can hear it again. My curiosity led me to hike towards the top of the mountain through the broken pathways lateral and longitudinal. When I reached the central height of the mountain the source of the sound came from my left side. As I hiked on the mountain's lateral path, I found myself facing the mountain opening and the source of the sound. Looking inside my eyes gauged the inner cave, slowly I took a step inside, penetrating through pitch dark

Realities Behind Vimana Aircraft

as I lose the light from the outside the mountain opening, the fear of disappearing in the dark cave became overwhelming. A sudden instinctive feeling froze me and prevented me to move further, so I analyzed from where I was standing. The sound came on and off but extremely loud, and I could not tell how far the object is and where exactly it was coming from. In one instance, my inner voice is telling that you have discovered what you have been looking for.

However, I found myself engaged in a dilemma, that any further step could potentially be dangerous, and it may not be the Vimana "the" flying chariot. It felt as if I am stuck in the time well, though, after regaining my senses and realizing that going back would be the best-case scenario, underlining, that all odds seem to be against me. So, I decided to turned around and found my way outside the mountain cave and followed the path to a nearby village. I walked for hours, and finally reached this small village, not the one I planned to reach but it seems safe and

Realities Behind Vimana Aircraft

welcoming. From distant, a man approached me who governed the village. We made an introduction, though, he didn't seem surprised and welcome me as he would welcome someone who he is close to him. While we were walking to the guest house that I could rent for, "Free", yes, I could not believe it either. When I offered money, he, refused and said it is not a norm here. I shared my experience with him and asked him if he can provide me a crew and portable lights to find out what was inside the mountain. The village caretaker explained it is not safe to travel after dark, therefore, it would be wise to travel in the morning. Meanwhile, he directed me to a guesthouse, where I can stay for the night and invited me to a dinner that his wife has prepared. I was tired, and respectfully decline the offer and accepted to take the food with me to the guesthouse. As the night went, I was excited with the thoughts that what I was looking for is finally discovered. As I closed my eyes, I woke up by the sound of morning hen, I quickly gathered my stuff and headed towards the village caretaker house. When I arrived at the village caretake house, as a courtesy he gave me lassi and bun for breakfast, which tasted delicious. There was also a band of men already waiting, prepared for the journey. The crew made an introduction and gave me some safety instruction before we began hiking early morning. The voyage was on the same path that I took a day before. I was losing confidence as we walked for hours, and covered much distance looking for the valley and the mountain where I discover the sound. For many hours, we travel across the valleys, and mountains, though, with no success on finding the lost valley I discovered a day before. We covered much distant and my strength seems to be weakening. One of the crew members suggests that if we don't turn back right now, we may not get back to the village before dark. I was not happy to return, so I argued to keep going, but villager men protested to return because moving forward could be deadly for everyone. I knew that the villager men were right; I just did not wanted to go

Realities Behind Vimana Aircraft

back to the States with the failure, especially when I am this close to discovery. In desperation, I ask the villager men to help me cover all directions at least for half a mile. So, we all divided North, East, South and West to find the valley that I found a day before and meet back at the same point. This way, I can plan another trip back the next day. It was unfortunate that none of us came back with any news. As I made it to the village, upon checking my calendar and money my flight back to JFK was due in four days, and I was running out of the money. So, I decided to end my journey until next time. After returning to Islamabad I shared my experiences with the people I knew, regarding the village I found so I can check out on google maps where exactly I was lost, though, none of them were able to point out the location nor the village name. I looked up a local map to reference all the villages and cities in the area that I perhaps encountered; I could not find any names referencing my location. So, I assume this village was either never was discover, or newly established. But the housing structure seemed as it was built from ancient times, houses look like Hindu and Buddhist temple. I did not have much time to research more on the location, but either way, I remembered the way to reach that remote village without a map. So, during my next visit, I will not return until I find the valley I discovered during my No-4 visit in Northern-Pakistan. I am confident that I will find the lost valley and will discover whatever was in that mountain. Do I think it is Vimana-ancient craft, yes, it is possible, or it is also possible that I accidentally discover something even greater? Regardless, where my next journey will take, I will continue my search for Vimana in northern Pakistan and the hidden areas around the Indus river.

A Look Into The Vimana Secrets

The researchers who look-into the mystery of UFOs often overlook an important fact. Though most of them assume that majority of the flying saucers have an alien origin,

Realities Behind Vimana Aircraft

there's another origin possible for these UFOs and that's ancient India and the Atlantis.

Everything that we know today about ancient flying vehicles of the Indian origin is available from the Indian sources of ancient times; these are the written texts which have reached us through these past centuries. It goes without a doubt that most of the texts that we're referring to are authentic. In fact, many of them are well-known Indian Epics from the ancient times. Out of hundreds of these Epics, many haven't even been interpreted in English and are still available in old Sanskrit.

In recent times, there was a news that a Vimana aircraft, the ancient Indian flying craft, was found somewhere in a cave in Afghanistan. However, there were different theories about it and many thinks that the aircraft was actually found somewhere in Pakistan. Regardless of where it is found, if such an object is spotted, there could be a lot of secrets that we may get to know about the existence of these crafts, the human alien interactions, the

Realities Behind Vimana Aircraft

lost cities in Pakistan, and the Atlantis. Let's try and discover more details about what the discovery of Vimanas could unravel.

The Vimana Mystery

The "Rama Empire" of India and Pakistan first developed somewhere around fifteen(15) thousand years back on the Indian sub-continent. This was a nation that boasted quite a few big and sophisticated cities. Some of these cities can be found in Pakistan's deserts, western, northern India and northern Pakistan even today. The Rama Empire existed parallel to Atlantean civilization and the "enlightened Priest-Kings" ruled this empire. This empire had 7 greatest capitals which are referred to as "The Seven Rishi Cities" in classical Hindu texts.

As mentioned in the ancient Hindu texts, these people had the latest technology such as flying vehicles that were known as "Vimanas". According to the Indian Epics, these Vimanas were circular, double-deck aircraft having a

Realities Behind Vimana Aircraft

dome and portholes. They had some serious resemblance to the flying saucers.

The ancient Hindu "Sanskrit" texts about Vimanas are so diversified that it would really take volumes for relating to what exactly they tried to convey. Ancient Indians manufactured these space crafts and they wrote complete flight manuals about how different kinds of Vimanas could be controlled. You can find many of these manuals even today and they've been translated to English as well. So, who knows, the discovery of Vimana might even reveal one such manual to human beings as well.

Only the popular Vimanika Shastra contains 8 chapters describing three different kinds of aircraft and explains them with diagrams. It highlights apparatuses that couldn't break or catch fire. It also describes 31 essential vehicle parts and 16 different materials that were used in their construction and can absorb heat and light. In fact, that was the very reason they were used in the construction of Vimanas.

Realities Behind Vimana Aircraft

According to the information shared in the Indian scriptures, the Vimanas had some type of "anti-gravity" mechanism to power them. They took-off vertically and they had the capability to hover in the sky.

So, if a Vimana is found, it could reveal a lot of amazing secrets that we – the modern-day humans – might be interested in. They could reveal the technology used at that time and will tell us about how advanced those civilizations were in this field.

The Atlanteans And The Vimanas

If you go through the Dronaparva section described in the Mahabharata, and Ramayana, it describes a vimana in a spherical shape. It depicts the movement of UFOs as it goes up and down, forwards and backward just as desired by its' pilot. Samar, another source of Indian origin, describes vimanas as smooth and well-knit iron machines with a mercury charge that shoots out of their back through that roaring flame.

Realities Behind Vimana Aircraft

Reading through these ancient Indian scriptures, it's quite evident that the ancient Indians used these vehicles to fly across Asia, to the Atlantis, and even to South America. There was writing discovered at the lost city of Mohenjo Daro in Pakistan – which is presumably among the "Seven Rishi Cities from the Rama Empire" and it resembled another undeciphered piece of writing found in the Easter Island. There is an uncanny similarity between the two scripts. Theorists say that Easter Island could have been the airbase in the Rama Empire that served the Vimana route.

To be unfortunate, as most of the scientific discoveries, Vimanas were used for a war ultimately. The Atlanteans used the flying machines called Vailixi – a known type of the aircraft – for literally trying and subjugating the world, according to the Indian scriptures referred to as the "Asvins" in comparison, the Atlanteans looked much more technologically advanced compared to the ancient Indians, because Atlanteans had more war-like

Realities Behind Vimana Aircraft

temperament. Even though we don't know about the existence of any texts about the Atlantean Vailixi, there has been a little information that came down through obscure, "occult" sources describing these flying machines. They may not have been the identical Vimanas, but they were rather similar and generally had a "cigar-shaped". They were also capable of maneuvering underwater and through the atmosphere reaching outer space. In similarity, some of the Vimana-like vehicles had saucer shape and had the ability to submerge under the water.

Eklal Kueshana, "The Ultimate Frontier" author, said in one of his 1966 articles, it was somewhere around 20000 years back when Vailixi's were developed for the first time in Atlantis. The most commonly made vehicles were the shape of a saucer and generally had trapezoidal cross-section, featuring 3 hemispherical engine pods to their underside. These vehicles used some mechanical antigravity equipment driven by their engines and developing around 80000^{hp}. Mahabharata, Ramayana, and various other Indian texts also talk about some hideous war between the Rama Empire and the Atlantis somewhere around 10000-12000 years back. They used destructive weapons to annihilate one another.

All these secrets could have been revealed by discovering ancient technology. It will validate historic events documented in Indian scriptures. However, the destruction of the lost cities of Harappa and Mohenjo Daro was also depicted in the ancient texts and this somehow attests to the event of some atomic war of the past.

The Unexplained Destruction of Harappa and Mohenjo Daro

Today we have the remains of the ancient cities of Harappa and Mohenjo Daro somewhere in Pakistan and India. These were the entire civilizations that vanished off the planet mysteriously. There hasn't been much explanation about their destruction to date.

Realities Behind Vimana Aircraft

The collapse of Mohenjo Daro is believed to be some 600 years by its inhabitance. Many archaeologists and theorists believe that this might have been the site where Indians and the Atlanteans must have fought their nuclear wars. Mohenjo Daro's excavations revealed skeletons lying around as if there was some abrupt event that must have caused their death. They might have been taken off by some unforeseen calamity. Some believe that a radioactive ash could reveal the source in the site's soil. All this points to some nuclear event featuring Vimanas and the ancient Indians. Though, we can't overlook the possibility of alien's interaction who could have staged these nuclear wars. Who knows!

Despite all the speculations, the claims of these two lost cities getting destructed due to some nuclear attacks are somewhat unsubstantiated. There isn't much evidence found about discovering any kind of radioactive materials at these sites. However, it is really strange that both these advanced cities had to face an apocalypse. What's even more strange is that the two cities were destroyed at the same time despite being located miles apart. The mystery of their demise is unsolved even today and the secrets might be unveiled with the discovery of vimanas and further studying the events that could have happened at that time.

The Possibility of Atomic Wars
In Mahabharata, there are many references that describe atomic war where vimanas were used to stage the destruction. Such references are not isolated; but the battles, using many different weapons and aircraft turn out to be a common aspect described in all the Indian Epics. One of them even describes a battle on Moon between a Vimana and a Vailixi craft.

As mentioned earlier, the excavation of Mohenjo Daro during last century found skeletons lying everywhere on the streets. A few of them had hands in hands suggesting that some sudden doom had stricken them. These, probably, are the most radioactive skeletons ever found, on par with the ones at Nagasaki and Hiroshima. In-addition, at Mohenjo Daro, a city that was very well-planned and laid on the grid and had plumbing system far superior to what we have today in India and Pakistan, various streets were discovered littered with the "black glass lumps". These glass globs were found to be the clay pots which must have been melted away under intensified heat which could be produced by nothing else than an atomic explosion.

As Atlantis sunk cataclysmically and the Rama empire was wiped out as a result of an atomic attack, the world seems to have collapsed into kind of a, "stone age", and it was only several thousand years after the event that our modern history would have picked up. Still, it seems that all the Vailixi and Vimana crafts of the Atlantis and Rama empire were not gone. They were constructed to last for several thousand years, and many of them might be used

Realities Behind Vimana Aircraft

even today, as proven by the "Nine Unknown Men" of Ashoka and the Lhasa scripture.

It is possible that these crafts and the scientific knowledge that would have been preserved by these societies of exceptional and enlightened beings. Many historical personalities including Buddha, Lao Tzu, Krishna, Confucius, Zoroaster, Quetzalcoatl, Mahavira, Akhenaton, Prophets and some recent inventors as well as other anonymous people could have been the line of enlightened beings.

An interesting fact is that at the time when India was invaded by Alexander the Great, his historians recorded their path of siege during the invasion was interrupted with an surprised attacked by a "flying, fiery shields" which frightened and plummeted his army . However, none of the "flying saucers" used any type of beam weapons or atomic bomb in this event, and they ended up conquering India.

Realities Behind Vimana Aircraft

Many writers believe that ancient societies used secret caverns somewhere in Tibet region and/or at some secret location in Central Asia to keep their Vailixi and Vimana crafts. Besides, there's a desert in Western China that goes by the name Lop Nor Desert and it is considered to have hosted some UFO mysteries as well. Probably, that's where many aircraft are kept even today, in some underground bases just the way Americans, Soviets, and the British do in recent times.

Despite that, all the UFO activity cannot be accounted for by these old Vimana crafts. It goes without a doubt that some of unexplained events are indeed governmental secret activities testing new technologies. Moreover, not all UFO sightings can be classified as swamp, clouds, gases, hallucinations, and hoaxes. There is a solid evidence available that many of these sightings, particularly "kidnappings" and similar nature events, were caused by "telepathic hypnosis".

Realities Behind Vimana Aircraft

I am not sure what is your take on it, but all these are the secrets that have been buried deep down in the history that may surface from time to time. Intentional confusion is created so one may turn away from the real truth. Holding on to any evidence even as small but reliable as a Vimana account may reveal many secrets that we're still not familiar with, knowledge is power, would you agree? knowledge is promoted by all religions, knowledge doesn't change the fact that only one higher power exists, "GOD", creator of all, from beginning of the time, before the garden of Eden, before the first man and before the first angel. Amen!

Realities Behind Vimana Aircraft

ABOUT THE AUTHOR

Salman Sadi is an experienced aviator, explorer, amateur astronomer, photographer, entrepreneur, and writer.

Salman Sadi is best known for exploring documented and undocumented historical landmarks around the world. His continued effort to discovering will be more promoted with the support of the audience.

Author: Salman Sadi | SalSadi010@gmail.com | SalmanSadi25@gmail.com

Link: https://www.linkedin.com/in/salmansadi

Made in the USA
Las Vegas, NV
28 January 2022